Gordon Bussey has been the Historical Consultant to Marconi plc (formerly GEC) since 1994 and has advised organizations across the world on the history of communications. He is the author, since 1976, of numerous publications on wireless-related subjects. Notably, he collaborated with Keith Geddes on the widely acclaimed book, *The Setmakers*. His recent works include *The first use of wireless by the Royal family 1898*, issued for a commemorative re-enactment in 1998, and *Marconi Centenaries in 1999*. He has been involved with exhibitions at national museums since 1977, and, most recently, with the Marconi section at the *BBC Experience*, Broadcasting House, in 1997.

In 1991 he was elected a Fellow of the Royal Historical Society.

To commemorate the centenary of Marconi's great achievement in bridging the Atlantic, the Royal Mint will be releasing a specially designed two-pound coin for circulation in the United Kingdom. During the preparation of this legal tender coin to be issued in 2001, Gordon Bussey has acted as historical consultant to the Royal Mint and its artists.

Front cover: Marconi at Signal Hill, see page 48

Back cover: Wireless masts at Poldhu, see page 33

Marconi's Atlantic Leap

Published by Marconi Communications 2000

First published in the United Kingdom by
Marconi Communications
New Century Park, Coventry
CV3 1HJ, England

Copyright © Gordon Bussey 2000

World copyright reserved

ISBN 0 95389 670 6

Designed by Aztec Design

Index by Paul Nash

Reproduced, printed and bound in the UK by
Cambridge University Press, Cambridge

Marconi's Atlantic Leap

GORDON BUSSEY

Acknowledgements

I would like to express my grateful thanks to Claire Blackman, Tom Going and Guy Peskett for reading the manuscript of this book, and for their helpful suggestions.

I have also received invaluable help from Debbie Beavis, Derek Burke, Rod Burman, John Carter, John Clarkson, Ian Forster, Bill Harrison, John Hooley, Dennis Knight, Ian Robertson, Bob Smallbone, Keith Smithers, Malcolm Walker, John Welford and Louise Weymouth.

I am indebted to Colin MacGregor for his generous help. He freely gave time and encouragement and made valuable suggestions.

I wish also to thank the following organizations: BBC Sound Archives, Cornish Studies Library, Helston Folk Museum, Helston Library, Institution of Mechanical Engineers Library, Morrab Library, National Maritime Museum London, The National Trust, Poldhu Amateur Radio Club, Provincial Archives of Newfoundland & Labrador, The Public Record Office, The Royal Archives, The Royal Institution and The Trevithick Trust.

Finally, I appreciate the opportunity afforded to me by Marconi plc of having access to the historical records which it has so carefully preserved over the years.

Gordon Bussey
Purley, February 2000

Contents

Marconi's Vision

From the beginning of his experiments, Marconi concentrated on increasing the range of his system. Many scientists believed that wireless waves travelled in straight lines from a transmitter and that they would not bend round the earth's surface, but Marconi was convinced that they would, and by the last years of the 19th Century he had shown that this was likely to be the case. In July 1897, he had established communication between the Italian cruiser *San Martino,* when its hull was below the horizon, and the shore station 10 miles away at La Spezia. In August 1898, he had shown that transmissions between the Royal yacht and Osborne House were weakened but not interrupted when hills intervened.

Still better results were obtained during the British Naval Manoeuvres in July–August 1899, when the best distance achieved was over 60 miles. An even greater distance was achieved in September 1899, in the Wimereux–Chelmsford tests, where the sea surface halfway between the stations was 1000 feet above a straight line between them. All this was enough to convince Marconi, if not all other scientists, that the curvature of the earth would not be a problem for much greater distances, and that to bridge the Atlantic, his next ambition, it was simply a matter of building a sufficiently powerful transmitter. However, Marconi was well aware that such a transmitter might cause problems:

"It was obvious to me from the very first," he wrote, "that if my purpose was to be brought to fruition it would be necessary to employ more powerful electrical waves than had ever been used before and above all things it was imperative that the production of these waves should not cripple or render useless the already established wireless communication between ships and the shore which was proving its enormous value more and more every day."

This was a problem that he solved at his experimental station at the Haven Hotel on the coast at Poole. This hotel was well-known to the yachting fraternity for its good French wines and cuisine. H M Dowsett (who joined the Company in 1899) quotes a visitor who was staying at the hotel in June

The Haven Hotel, 1898. Marconi established here a wireless station in October for his early transmissions. He kept it for experimental work until 1926

of the same year and described what he saw. Just outside the boundary walls there was the 110-feet stepped three-piece hemp-guyed Marconi mast, with its stay anchors under water at high tide. There were huts in the grounds for wireless stores, but the main laboratory was a large room on the ground floor. The aerial wire led in through the centre of a plate glass window to the coherer receiver and induction coil apparatus, used for communication with Marconi's wireless station at the Royal Needles Hotel on the Isle of Wight (IOW).

During the greater part of the day, signals were being received continuously for testing purposes, and both stations, in order to demonstrate their good working condition, had to receive and repeat back Morse code messages first thing every morning, atmospherics or no atmospherics, and the Morse inker tape records were sent up to London to be checked. At one table were the brothers J and R Cave making coherers (detectors of wireless signals), at another was P W Paget winding receiver chokes in the intervals between making experimental receiving jiggers (transformers). Marconi himself was at his own table occupied with fitting V-gap plugs into an experimental coherer. Nearby was another table with a 500-volt battery used by Marconi for tests with experimental glow discharge coherers. Outside, along the foreshore, Dr Erskine Murray and two assistants were conducting tests at centimetre wavelengths using parabolic mirror reflectors.

Incidentally, this visitor gives a rare account of Marconi actually relaxing. At meal times, Marconi, his mother and sometimes Alfonso his brother, Dr and Mrs Erskine Murray, the occasional guests of the Company, and the Marconi staff "all shared a common table where general topics and the day's results were discussed in an atmosphere that was delightfully informal yet dignified. Often on those quiet evenings when relaxation after working hours was sought at this quiet spot, Murray would bring out his cello, Alfonso the violin and Marconi would accompany them on the piano."

The importance of the Haven Hotel for future transmissions from the

One of Marconi's hotel bills, 1901

Poldhu station was that it was here that Marconi perfected his so-called syntonic system of selective tuning, in which other circuits in the transmitter and receiver, as well as the transmitting and receiving aerials, were tuned to the same wavelength. This was an advance on Sir Oliver Lodge's idea, patented in 1897, of only tuning the sending and receiving aerials. Marconi had made it possible for multi-station operation without mutual interference. He incorporated his techniques for achieving this system of tuning in his famous patent No. 7777 of April 1900.

By the middle of 1900 Marconi was convinced that it would be practical to send messages across the Atlantic, but he was faced with the great problem of raising enough money to construct the powerful transmitter needed. The ship-to-shore business was not yet fully established and his Company was living on its capital. His arguments for experimental expenditure would have to be very convincing to win the other Directors' approval. Meanwhile, he went to the United States to demonstrate his new inventions and to obtain American patent rights. On the 2nd June,

The SS Teutonic *in which Marconi returned to England in June 1900. Liverpool is in the background*

accompanied by W Densham (one of his assistants) and with cases of apparatus, he disembarked from the American Line SS *St Paul* at New York. He had interviews with the Marconi Company's agents, Messrs Moeran and Bottomley, and several interesting talks with his good friends in the newspaper world, particularly the New York Herald.

By the time that Marconi landed back at Liverpool on board the White Star Line SS *Teutonic* on the 28th June he had made up his mind that the transatlantic test must be put in hand without further delay. Having reported to the Board after his American visit, he gave a demonstration of selective tuning between the Haven and Niton (another experimental station on the IOW) to the Managing Director, Major S Flood-Page. This convinced the Board that his new syntonic tuning patent could be used to prevent their ships' wireless telegraphy business being interfered with by powerful transmissions from the proposed experimental plant. Towards the end of July, the Board agreed to the transatlantic test.

Building Poldhu

Marconi, with Major Flood-Page and R N Vyvyan, an engineer recently appointed, set out to select a suitable site for the new station. The Cornish coast offered advantages and, early in August, Marconi visited and gave his final approval to a site on Angrouse Cliff overlooking Poldhu Cove, the lease of which was acquired from Viscount Clifton. Plans for the buildings were prepared by the Marconi Company in London, and work on the site, supervised by Vyvyan, started in October.

Dr J A Fleming (later Sir Ambrose Fleming, FRS) of University College, London, had, since his letter to *The Times* describing Marconi's cross-channel achievement in 1899, kept in touch with all the latest developments. Fleming was appointed consultant to the Marconi Company on the 1st December 1900, for three years at a salary of £500 per annum. However, he had been assisting Marconi before this on the design and construction of the electrical engineering plant required for the generation and control of the powerful electromagnetic oscillations (on a short-term consultancy basis at a lower salary). The high-frequency and aerial circuits were being devised by Marconi himself.

Fleming was closely associated with Marconi during this period and wrote:

> "Up to that time Marconi had only used laboratory apparatus,
> such as spark induction coils, Leyden Jars, and simple

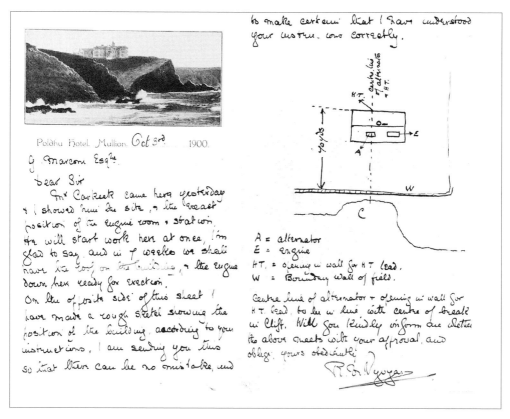

Report by Vyvyan to Marconi from Poldhu, October 1900, confirming their plans for building the wireless station

transformers. Long-distance wireless involved the transformation of this laboratory apparatus into engineering plant. The Marconi Company had appointed me their Scientific Adviser and I was asked to specify the required Power Plant. A site was selected in Cornwall at Poldhu, near Mullion, and my first work consisted in designing the required building and machinery."

During the many months of building the Poldhu station, Marconi was constantly travelling between it and the Haven Hotel. On arriving at Parkstone Station (2½ miles from the hotel) it was a long train journey with changes at Temple Combe Junction and Exeter and then on by the Great Western line to Helston. From here it was 7 miles by road to his destination

Map of the Cornish coast showing the Poldhu Hotel and the layout of Marconi's wireless station that had been completed by July 1903. This appears to be the first official map showing the wireless site

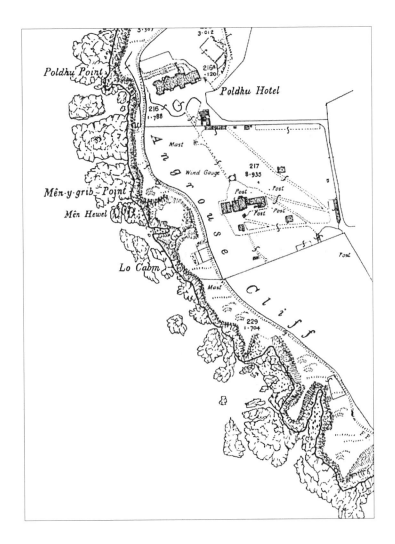

and he would be met by the horse-bus that the Poldhu Hotel provided for its guests. Alternatively, he would go through London for the purpose of attending to Company business on the way. The railway journey from Paddington to Poldhu would involve Marconi leaving a Penzance train at Gwinear Road, where he would change for Helston.

Marconi realised that he needed an additional wireless station within easy reach of Poldhu. A suitable site was found 6 miles away at the Lizard (the station at Niton on the IOW was too far away). The new station would provide:

17

1. another installation for ship-to-shore traffic;
2. local facilities for testing Poldhu transmissions;
3. a nearby site for checking newly devised receivers. It was feared that these might suffer interference from a new station as powerful as Poldhu.

A Marconi station at Dovercourt was closed on the 27th December 1900. Its three-piece mast, with a total height of 161 feet, was lowered by George Kemp, Marconi's assistant, and despatched to the Lizard where it was re-erected by him, with the assistance of a rigger called Groves, early in January 1901. Instructions to build a hut to hold the apparatus were given to George and Sons, a local builder, and this was completed by the 18th January. The first signals on the completed installation were received from Niton by Marconi on the 23rd January. This established a new distance record for receiving syntonic tuned transmission of 186 miles. All other coastal stations in the Channel, the Haven, Portsmouth, Portland, Plymouth and on the French coast, were picked up during the next few days, Paget assisting with the production of suitable receiving jiggers. Kemp's diary records that in June Paget was in charge of the Lizard station and resided at the Housel Bay Hotel.

The original plant at Poldhu consisted of a 32-brake-horsepower Hornsby-Ackroyd oil engine (about half the power of that fitted in a modern small car) driving a 25-kilowatt 2000-volt Mather & Platt 50-cycle alternator. The output was fed to a pair of 20-kilowatt Berry transformers which stepped up the voltage from 2000 volts to 20,000 volts and this was then applied to a bank of condensers discharged via a spark-gap into two coupled HF transformers.* This arrangement gave the very high oscillatory energy which Marconi had calculated was needed to get signals across the Atlantic. The final output, 100 times more powerful than earlier

* See Fleming, *The Principles of Electric Wave Telegraphy*, pages 503/505. Longmans, Green, and Co. London, 1906.

Key to the gate between the wireless station and the grounds of the
Poldhu Hotel, ca. 1903. Note the label written by Kemp

transmitters, was inductively coupled to the aerial. Fleming developed an ingenious way of keying the transmitter in the 2000-volt circuit which avoided arcing problems.

Early in February, with the power plant equipment under test, Vyvyan left Poldhu to make all the necessary arrangements in London for the construction of a similar station on the other side of the Atlantic. Poldhu came under the temporary charge of C H Taylor.

Details of the proposed ring of masts to support an inverted cone-shaped aerial arrived from London on the 12th February. Twenty four-piece 200-feet masts (lowermast, topmast, topgallant and Royal mast) were to be set out in a circle 200 feet in diameter, and Kemp started work on this at once.

Key to the gate between the wireless station and the cliff, ca. 1903.
Note the label written by Kemp

He purchased materials from suppliers locally and from London. Some five weeks later, E A N Pochin who, since the previous December, had held the position of Chief Engineer in the Marconi Company, inspected the progress of the station pending the return of Marconi, and discussed the mast assembly and its problems with Kemp.

Kemp worked six days a week on the aerial masts from this time until November. This is a good moment to stress how much Marconi owed to this devoted man. He had assisted Marconi from 1896 (being officially employed by the Marconi Company in 1897) and stayed until he, himself, died in January 1933. His devotion is illustrated in two incidents recorded in his diary. In April of 1901 he was recalled to Chelmsford for a few days and he describes his return journey: "I left here (Chelmsford) at 2.30 a.m.

Horse-bus from the Poldhu Hotel meeting the train at Helston, ca. 1903

and arrived at Liverpool Street at 3.45 a.m. I walked through London to Paddington in 2¼ hours and, by 4.30 a.m., it was daylight. I left Paddington at 8.00 a.m. and arrived at Guinea (sic) Road at 6.30 p.m. and, from here, I travelled to Helston and on to *The Poldhu Hotel*." Then in September he records how he helped Marconi personally: "I assisted Mr. Marconi to prepare his new front driven motor cycle, after which the remainder of the Sunday was spent in making up accounts and correspondence."

The new Engineer-in-charge at Poldhu, W S Entwistle, who had been appointed to the staff early in the year, arrived to take over his duties on the 31st March 1901. Two other special staff appointments were made at this time, both on the 1st April 1901. H Cuthbert Hall was made Manager of the Marconi Company to take over much of the work up till then undertaken by Major Flood-Page, and Captain C V Daly was made Marine Superintendent of the Marconi International Marine Communication Company formed a year earlier. The latter's wide experience

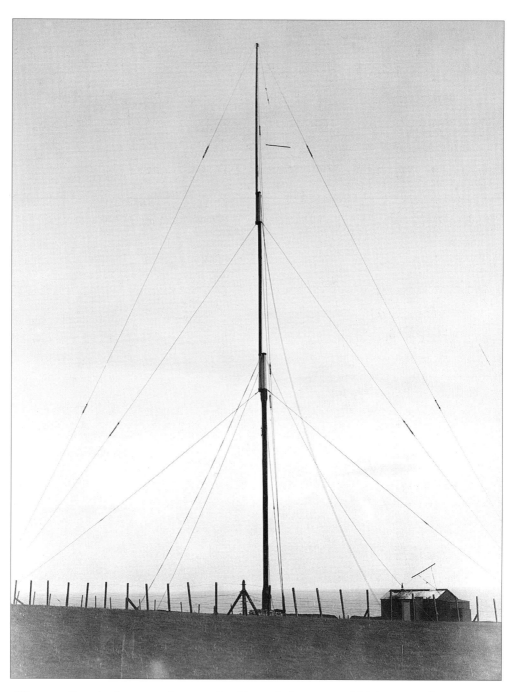

Marconi's wireless station at the Lizard, January 1901, recently completed

of conditions on the North Atlantic traffic routes enabled him to obtain useful contacts both with the men who commanded the big ships as well as with their owners.

Marconi, accompanied by Vyvyan, left England for the United States at the beginning of March and chose a site for the American Transatlantic station among the sand dunes of South Wellfleet, Cape Cod, Massachusetts. Then he left Vyvyan to undertake its construction and returned to England by the Cunard Line SS *Campania* at the end of April. This was in time to prepare his paper *Syntonic Wireless Telegraphy* which was read at the Society of Arts (Royal in 1908) in London on the 15th May.

Experimental work of various kinds in connection with the power-control and high-frequency circuits had been going on at Poldhu since January under the direction of Marconi and Fleming. After receiving Poldhu experimental transmissions at Niton, a distance of 186 miles, Marconi, accompanied by P J Woodward, took his so-called long-wave receiving apparatus to Crookhaven in County Cork towards the end of June and checked the strength of the Poldhu signals. Crookhaven, 225 miles from Poldhu, had been opened at the beginning of June with Frank Newman in charge. Morse code dots were being sent by Densham and Kemp at Poldhu on the modicum of power then available at an estimated wavelength, approximately 366 metres. Marconi paid two further visits in the second half of July, and the results over this greater sector of the earth's curvature fully supported Marconi's belief in the ultimate success of the transatlantic venture. Towards the end of the year Newman was joined at Crookhaven by A H Ginman.

The erection of the masts at Poldhu proved difficult, due partly to the heavy weather and partly to the nature of the construction. At one stage in August the thunder caused the horses to stampede and the men to leave the 10-acre enclosure in great haste. To reduce currents induced in the wire stays, the stays were broken into sections by dead-eye and lanyard inserts

continued on page 35

Above: *Interior of the Lizard wireless station showing Marconi's apparatus ca. 1903. The picture on the wall is of the American Line ship SS* St Louis

Opposite top: *The wireless station at Poldhu April 1901, showing the brick-built power house with the first experimental aerial. A zinc cylinder hangs from a pole which has been hoisted to the top of the main construction derrick. In the foreground are the poles for the main aerial*

Opposite bottom: *The power house, seen from the other side*

Inside the Poldhu power house, April 1901

The alternator and oil engine in their operational location. Note the drive belt

Opposite above: *The Hornsby-Ackroyd oil engine of 32-brake-horsepower which drove the alternator*
Opposite below: *The Mather & Platt 50-cycle alternator which provided the high-voltage transformers with 25-kilowatts at 2000 volts*

Part of the apparatus used for the first experimental transmissions.
Left to right: two choking coils, two high-voltage transformers, the main
condensers and, in front of the window, the spark-gap used at this time.
This 'double deck' arrangement of condensers proved unsatisfactory,
and they were moved to the wooden transmission hut in August 1901
(untouched original photograph)

Opposite: *The secondary jigger (1901 description) shown in two*
different stages of operation. It was moved to the wooden transmission
hut in August 1901

Poldhu site, April 1901. The first lowermast is on the left, erected with the aid of the derrick seen on the right

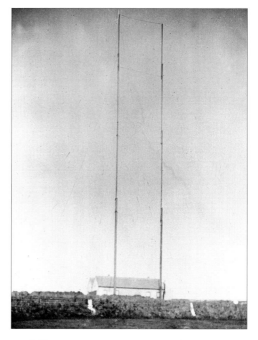

Poldhu site, May 1901. The first two of the twenty 200-feet masts are shown complete

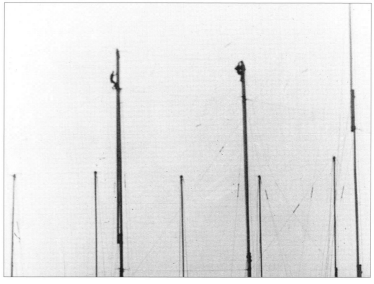

Poldhu aerial nearing completion in August 1901. Note the two riggers preparing to fit further 'Royal masts'

Above: *Marconi's Poldhu wireless station in its nearly complete form, August 1901. The oil engine, alternator, high-voltage transformers and choking coils were in the brick-built power house, seen on the left. The condensers, primary and secondary jiggers, tuning inductance and spark-gaps were in the wooden transmission hut seen on the right. The aerial wire entered at the exact centre of the roof of the hut*

Opposite: *Poldhu Hotel seen in relation to the almost complete circular aerial, August 1901*

and only the radial stays were taken to anchor blocks, the circumferential staying being obtained by means of horizontal triatics taken from one mast to the next. There was the risk that a fault in one of these triatics could endanger the security of the whole assembly of 20 masts. This risk had been discussed but it was accepted on account of the diminished absorption by these triatics which were at right angles to the wires of the aerial within the circle of the masts.

Actually, the test came before all the masts were completely erected and while Marconi was working with the only section of the aerial in service. On the 17th September a south-westerly gale suddenly swung round to the north-west. A heavy squall struck the Poldhu masts, a lug was torn from one of the top gallant mast caps and the whole circle was wrecked.

The Marconi Company was appalled by this reverse. It meant a heavy capital loss at a period of financial stringency, and also delays and doubts about the design of the Cape Cod aerial. However, construction was so far advanced that a decision was taken to continue without modification.

The day following the crash at Poldhu, Marconi left for London while Kemp and his gang began to clear the tangle of masts and stays and to re-erect a serviceable structure as soon as possible. An experimental aerial was available on the 24th and Marconi resumed his tests on the 26th. Work was again in full swing when, during the next fortnight, individual members of the Board, H Jameson-Davis, Major Flood-Page, J F Bannatyne and Cuthbert Hall called at the station to view the damage. At the Lizard they were also shown long-distance reception from Niton.

On the 30th September Marconi left Poldhu to meet his friends of the Italian Government Wireless Commission in London, and then he proceeded immediately to Niton, the Haven and later the Lizard. Here he carried out exhaustive tests of telephone receiver reception with

Opposite: *Scene of devastation, 17th September 1901. The Poldhu circular aerial after the gale*

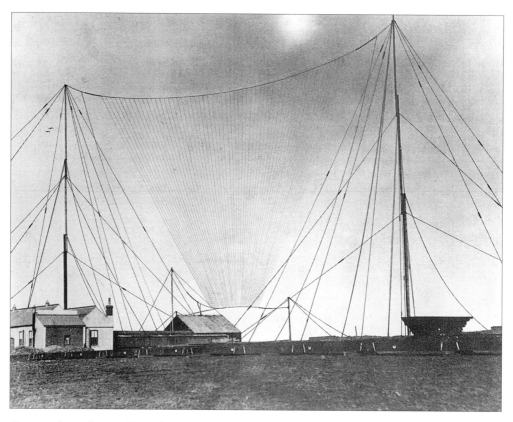

Seven days later, Tuesday the 24th September 1901: the temporary aerial is erected and used by Kemp

a self-recovering mercury coherer which his companion Lieut. Solari of the Italian Navy had brought with him (Marconi had been conducting experiments with a similar coherer at Poldhu from the 8th September). There does exist Marconi's authority for calling this coherer the 'Italian Navy Coherer', as he quotes a letter dated the 4th June 1902, from the Head of the Italian Navy, saying it had been the fruit of the work of various individuals and not just of one person.

A proposal by a Marconi engineer, A E Heming, to design and erect four wooden towers, 200 feet high, at the corners of a square of 200 feet side-length, in place of the collapsed 20 masts, was considered at Poldhu. The use of wood for such a height, windage and expected masthead load

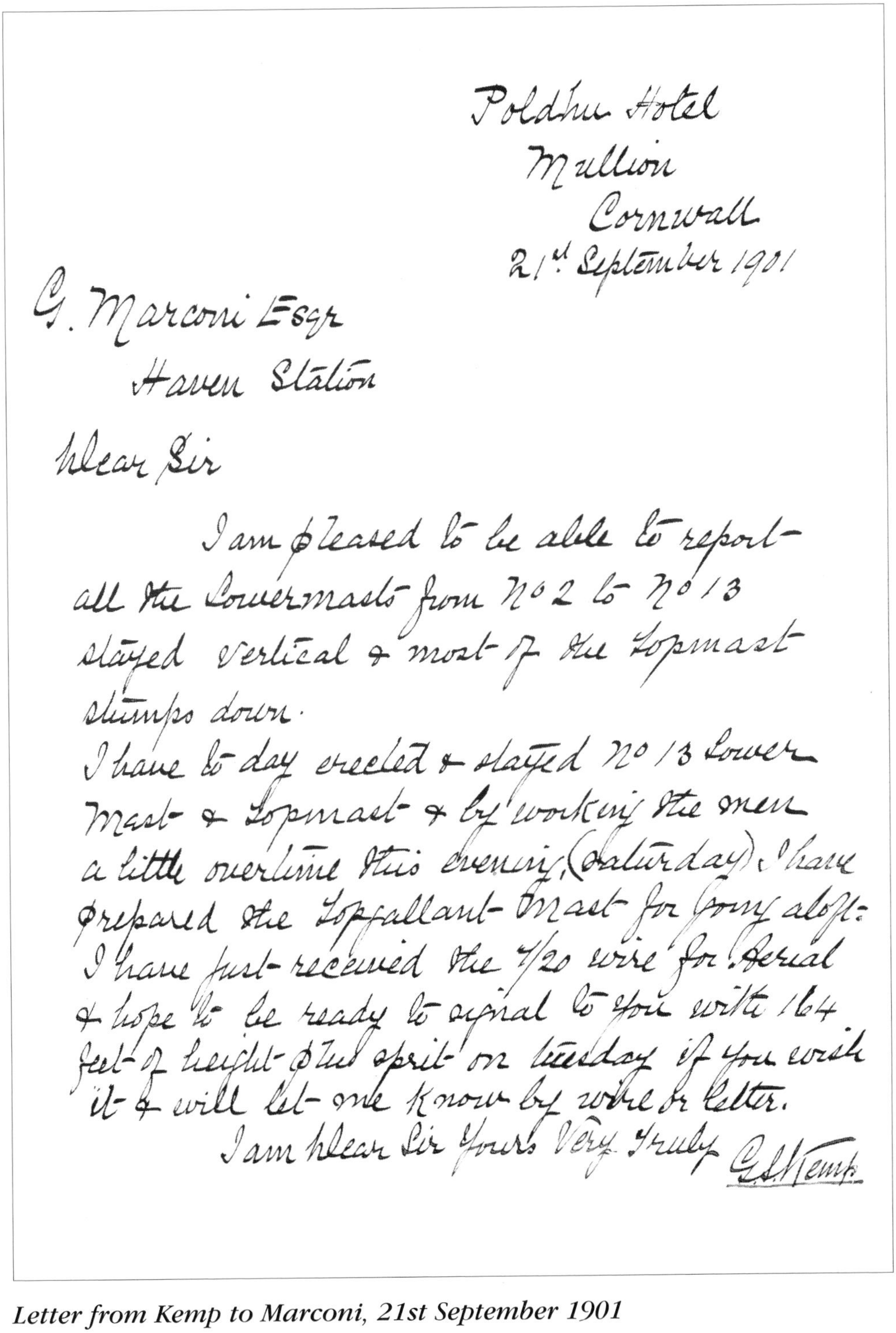

Poldhu Hotel
Mullion
Cornwall
21st September 1901

G. Marconi Esqr
Haven Station

Dear Sir

I am pleased to be able to report
all the Lowermasts from No 2 to No 13
stayed vertical & most of the Topmast
stumps down.

I have to day erected & stayed No 13 Lower
Mast & Topmast & by working the men
a little overtime this evening (Saturday) I have
prepared the Topgallant Mast for going aloft.
I have just received the 7/20 wire for Aerial
& hope to be ready to signal to you with 164
feet of height & the sprit on tuesday if you wish
it & will let one know by wire or letter.

I am Dear Sir Yours very truly
G.S. Kemp

Letter from Kemp to Marconi, 21st September 1901

was an untried novelty. Heming's stress diagrams, however, when checked over by C E Rickard (a future Chief Engineer), were shown to provide a reasonable safety factor, and on the 22nd October Marconi finally agreed that Heming should proceed with the construction. Work on the foundations began on the 1st November.

During this period Pochin, who had been spending a lot of time at Poldhu, transferred his activities to the Chelmsford Works. Subsequently, he resigned from the Company and founded his own business. Andrew Gray, on returning from the United States, took over Pochin's duties with the title (at Marconi's suggestion) of Chief of Staff. The title reverted to Chief Engineer some years later.

Marconi's progress in extending the range of signalling had up to this period been methodical and scientific, and every increase had been checked and consolidated by repeated tests and improvements in the apparatus. Now, however, with promisingly strong signals at Crookhaven from the Poldhu plant using an experimental aerial, he felt that nothing more should be done to the transmitter until a receiving test had been made across the Atlantic, even though it must incur the risk of failure. So he decided that he would try to send a message to the nearest point on the American continent rather than try to get to Cape Cod, and that meant sending it to Newfoundland (a British Colony then but a province of Canada later). The Board agreed and felt it would relieve the present suspense caused by the uncertainty that transmissions would reach as far as Cape Cod.

The aerial system at Cape Cod, USA, Autumn 1901. It was the counterpart to the Poldhu aerial and was also destroyed by a gale

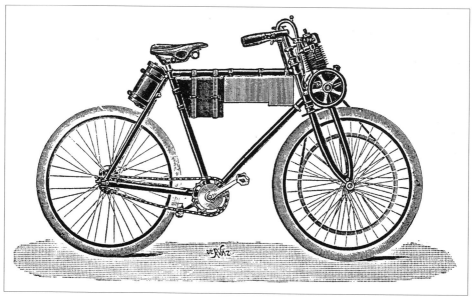

A preliminary list, by Marconi, giving items required for his wireless receiving station in Newfoundland, October 1901

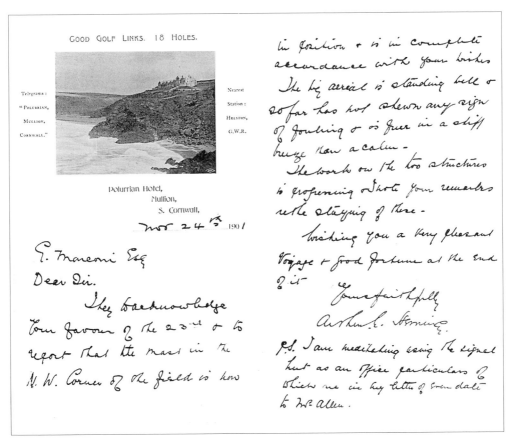

Letter from Heming to Marconi reporting on progress at Poldhu,
24th November 1901

Opposite: *A Werner motor bicycle with a small*
(1 nominal HP) motor of the type bought by
Marconi in September 1901 from the Motor
Manufacturing Company, Holborn, London,
and which Kemp helped him prepare at Poldhu

The SS Sardinian *on which Marconi, Kemp and Paget travelled to Newfoundland on the 26th November 1901*

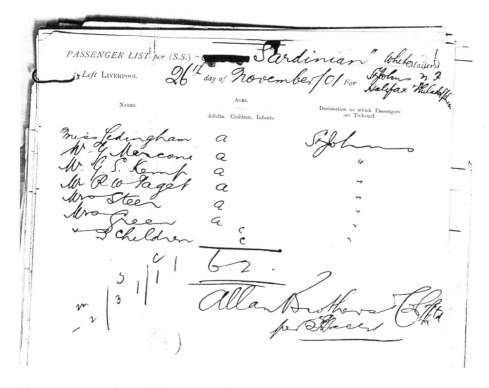

The passenger list of the SS Sardinian

Triumph at Newfoundland

On the 26th November 1901, Marconi, accompanied by Paget and Kemp to assist in the work, embarked on the Allan Line SS *Sardinian* for Newfoundland. Ostensibly they were going in connection with a ship-to-shore wireless enterprise in that part of the world about which the Company had had an enquiry. However, their equipment included hydrogen gas cylinders to inflate cotton-silk balloons, 14 feet in diameter and having a capacity of 1400 cubic feet, six Baden-Powell kites that measured 9 feet by 7 feet, aerial wire, and receiving apparatus. Just before the ship sailed, Captain Daly (Marine Superintendent of the Marconi International Marine Communication Company) arrived with a telegram giving the news that the masts at Cape Cod had suffered the same fate as had befallen those at Poldhu.

Landing at Shea's Wharf, St John's, Newfoundland on the 6th December, Marconi was met on behalf of the Government by the Minister of Marine and Fisheries and the Inspector of Lighthouses, who offered him every facility in the carrying out of his work. Signal Hill, a rocky eminence near the mouth of the harbour and overlooking the town, proved in most ways to be an excellent site for kite-supported aerial reception, though later Marconi was to say that the earth connection was not good owing to the very rocky condition of the sub-stratum. The authorities made available two rooms in a disused fever hospital attached to a military barracks on

43

Passengers on board the SS Sardinian. *Left to right: Kemp, Marconi, boy (probably son of a Mrs Green) and Paget*

the hill. Outside there was sufficient room on the crown of the hill to fly kites or balloons.

On Monday, the 9th December, an earth connection was made with buried zinc plates; two kites were prepared with their lines and the No. 2 balloon was oiled and made ready for inflation. The instruments were set out and wired up on a table in a lower room of the hospital and the aerial wire passed through a window to a telephone pole, from where it was attached to a balloon or kite. Entwistle at Poldhu was cabled via the office in London to start the pre-arranged programme of transmissions from the 11th.

The great moment would soon arrive. Marconi, keenly sensitive to historic connections, found that the site chosen for a receiving station was close to

Premises at Signal Hill, Newfoundland, used by Marconi to house a wireless receiving station, December 1901. Note the aerial held aloft by a kite

Signal Hill, December 1901. Left to right: Kemp, Marconi and Paget are shown with a wicker basket used to contain apparatus. Note the size of the erected kite seen behind them

the memorial tower commemorating the discovery of Newfoundland by John Cabot at the end of the fifteenth century, and thus establishing sea communication between that outpost of the Western Hemisphere and Europe. Not so far distant, too, was Hearts Content where the first successful transatlantic cable was landed in 1866, thereby effecting another form of communication by submarine telegraphy with Europe. Now it was Marconi's turn to show at the same outpost, if his intuition was not at fault, that a third form of communication, by wireless telegraphy, was not only possible but a reality.

On Tuesday, the 10th December, one of the aerial-support kites was given a trial flight. On the day after, Wednesday, the 11th December,

An attempt to raise the kite-supported aerial at Signal Hill, December 1901. Note that the lower picture shows Marconi on the left

Marconi pictured at Signal Hill, Newfoundland just after receiving the
first wireless signal to cross the Atlantic on the 12th December 1901.
It consisted of the Morse code letter 'S' (dot dot dot) transmitted from
his wireless station at Poldhu

the weather was fair but the wind soon started to freshen. Poldhu had been instructed to transmit the Morse code letter 'S' (dot, dot, dot) daily from 3.00pm to 6.00pm GMT (11.30am to 2.30pm, St John's local time). The three-dot letter 'S' was chosen because the switching arrangements at Poldhu on that occasion were unable to withstand the wear that a long series of dashes would have involved. Also, it ensured that the signal could be read on a Morse inker recorder through atmospherics which tended to run all the dots and dashes into each other. Under these conditions a series of three dots ('S') was more distinguishable than a series of any letter containing dashes. Similar reasoning applied to aural reception.

Fortunately Kemp's diary gives an account of what happened on these vital days in December:

11th: "The balloon was partly filled with hydrogen gas and

I had great trouble with it. This was due to the strong wind that had suddenly sprung up at 3.00 p.m., which carried away the special mooring line, supplied by Mr Spencer, and, had it been carried away on the mooring side instead of the balloon side, I should have gone with it as its speed was like a shot out of a gun. Mr Marconi tried all the detectors from time to time, while we had the balloon under control, and called me in at intervals to try the receivers. Signals appeared at intervals on a telephone in series, when using our sensitive tube circuit, and, at times, the dots threatened to appear on the tapper."

It is clear from this that some signals were received but apparently were not definitive ones. Unfortunately, Kemp does not record any details of the circuits or coherers used, but the mention of a tapper suggests that at least one of these was a filings coherer. On the next day Kemp describes the great triumph:

12th: "We received the three dots or the S signal repeated. We lost the first kite with 2 wires, each 510 ft. long, after it had been the means of giving us one hour for reception which was better than yesterday. Another kite was raised with one wire 500 ft. long, which appeared more in harmony with the earth's electric medium and the signals from Poldhu Station. We were able to keep this kite up for three hours and it appeared to give good signals."

Marconi himself was listening and described this as the crowning moment of his life. He wrote:

"Unmistakably, the three sharp little clicks corresponding to three dots, sounded several times in my ear, but I would not be satisfied without corroboration."

"'Can you hear anything, Mr Kemp?' I said, handing the telephone to my assistant."

"Kemp heard the same thing as I and I knew then that I had been absolutely right in my calculations. The electric waves which were being sent out from Poldhu had traversed the Atlantic, serenely ignoring the curvature of the earth which so many doubters considered would be a fatal obstacle, and they were now affecting my receiver in Newfoundland."

Marconi himself gave further details of what happened in his statement on the 23rd December to the Chairman and Directors of his Wireless Telegraph Company:

"...On Thursday the 12th., a successful attempt was made to rise (sic) one aerial by means of a kite, and at 12.30, 1.10, and 2.20, the pre-arranged signals from Poldhu were received in a manner which left no room to doubt their authenticity by myself and Mr. Kemp, on the telephone receiver, but not on the ordinary receiver. Signals were also received, but less distinctly, on Friday the 13th. On Saturday the 14th., a strong gale from the northwest made it impossible to elevate either kites or balloons."

Marconi had provided for a third possible observer, but the Wednesday session did not last long enough, and on the Thursday Paget was indisposed. On Saturday the 14th, when the weather continued foul, Marconi had to call a temporary halt to his tests. Some alternative method of supporting the aerial had to be found. The sections of the mast to be erected at Cape Race in connection with the proposed service to ships operating off the Grand Banks (which stretched for some 300 miles south-east into the Atlantic) had just been inspected and approved for erection. Could this be used? Kemp and his two men, from Newfoundland, during the morning fitted a wire down the cliff from Signal Hill to an iceberg stranded in the harbour, so providing another form of temporary aerial that could be tried out.

On Thursday, the 12th, Marconi had drafted a cable to Major Flood-Page reporting the success of his tests, but on second thoughts he withheld it in the hope that the weather might improve and provide better test conditions before its despatch. In this he was disappointed as the weather worsened. Though the signals on the 12th had convinced Marconi, he realised that they constituted slender evidence to an impartial critic upon which to proclaim the conquest of 1800 miles of ocean by wireless. The alternative explanation of what had been heard, atmospheric discharges, was bound to be suggested. If this was accepted it would reflect on the prestige of his Company and might affect its future business. However, the need to keep the Company informed, and his inability to carry out further tests quickly, compelled Marconi to send the cable and take the risk (cables were not private).

The cable was addressed to the Managing Director, Major Flood-Page at London. Having written it on the 12th it was despatched on the 14th. It read as follows:

> "St John's, Newfoundland, Saturday, 14 December, 1901. Signals
> are being received. Weather makes continuous tests very difficult.
> One balloon carried away yesterday." (The "yesterday" must refer
> to the 11th if written on the 12th.)

Speaking about his triumph in a later sound recording, Marconi himself said:

> "The experiment had involved risking at least £50,000 to achieve
> a result which had been declared impossible by some of the
> principal mathematicians of the time."

Marconi pictured with the Governor and Members of the Newfoundland Government at the Cabot memorial tower, Signal Hill, December 1901

Public Reaction

It is not surprising that the Press was sceptical at the news. Marconi was making such an extraordinary claim. In an unsympathetic report *The Daily Telegraph,* on the 18th December 1901, rejected the tests as inconclusive:

> "The view generally held is that the 'electric strays' were responsible for activating the delicate instrument recording the 'S's' supposed to have been transmitted from near the Lizard to Newfoundland... Some attribute the wandering currents to the old trouble – earth currents. Others to the presence of a Cunarder fitted with the Marconi apparatus, which was, or should have been, within 200 miles of the receiving station at St John's on the day of the experiment."

The Times was slightly more favourable. In an editorial of its issue on the 21st December 1901, it wrote:

> "There seems to be something far more than a probability that the close of the first year of the century has witnessed the greatest triumph of applied science in the department of electricity which has so far rewarded those who have selected this department as the department of their labours.
> Mr Marconi has gradually accustomed us to the wonders

of wireless telegraphy, so that we have come to hear with but mild surprise of the transmission of his signals over distances of one mile, of ten miles, of fifty miles, and so forth; but, even with this experience, it is in some degree of shock to all preconceived notions to be told that he has received on the shores of Newfoundland signals transmitted from his station at Poldhu in Cornwall, and that he even thinks it possible for wireless telegraphy across the Atlantic to be sufficiently developed within the space of four years as to fulfil all ordinary commercial requirements." (A full commercial transatlantic service was in operation by February 1908.)

The technical press was no less sceptical. In its issue of the 17th January 1902, in the course of its leading article, *The Electrician*, not denying that signals had, in fact, been transmitted across the Atlantic, attacked the feat on the score of technical obstacles that stood between it and any practical application for commerce. A service of 50 words a minute, it argued, would make necessary ten gigantic stations on both sides of the Atlantic, with probable interference with all other signalling, and a probable tapping from the Urals to Spitzbergen and the Gold Coast. And both the lay and technical Press also argued that to perform this feat, aerials 200 feet high would be essential.

But there was another criterion by which the Company was able to measure the reaction to its sensational success, namely that it was taken seriously by those who represented the great vested interests of the cable companies and land telegraph systems as is shown by a letter that Marconi almost immediately received from the Anglo-American Telegraph Company.

Letter to Marconi from Sir Robert Bond, the Premier of Newfoundland, congratulating him on his achievement, 14th December 1901

"The Grange",
Whitbourne.

Dec 14. 1901.

Dear Mr. Marconi :-

On my arrival here
this morning I received a
telegram from the Minister
of Marine & Fisheries informing
me of your great success in
communicating a distance
of Eighteen hundred miles.
Permit me to Extend to
you my hearty congratulations
on your wonderful Triumph.

Sincerely Yours

W. Coad

G. Marconi, Esqr

On Sunday, the 15th December, a day of rain and fog, having attended divine service, Marconi and his small party were guests at a quiet celebration. On the 16th the news was given a general release. Mr R White, the Inspector of Lighthouses, drove out with Kemp during a blizzard to inspect an alternative site for an experimental station at Cape Spear and returned in dense fog. But that very evening Marconi received a letter from the solicitors of the Anglo-American Telegraph Company, over whose lines the cable to London on the Saturday had been sent, calling his attention to the monopolistic rights with which the Company had been invested by its Charter, one that precluded any sort of competition in or from the Colony. Under threat of immediate legal proceedings, Marconi was called on to discontinue his work and to dismantle his station.

As a result of this threat Marconi cancelled immediately all preparations for further tests, but on the following day, the 17th, gave a working demonstration of his apparatus on Signal Hill to the Governor and Executive Officers as previously arranged. He also sent this cable to London:

> "In consequence of the threatened legal proceedings against
> myself by the Anglo-American Telegraph Company which
> claims a monopoly of telegraphic communication in
> Newfoundland, I am compelled to discontinue wireless
> telegraph tests between England and this Colony. I hope to
> continue shortly with similar experiments between England,
> Canada and the U.S.A." (The last two countries were free of
> monopolistic rights.)

His legal adviser in Newfoundland was Sir James Winter KC, and Marconi instructed him to send the following letter to the Anglo-American Telegraph Company's solicitors. He reported this to the Directors of his Company.

St John's, Newfoundland
Dec. 18th, 1901

Messrs. Morine & Gibbs,

Dear Sirs,

I am instructed by Signor Marconi to Acknowledge receipt of your last letter to him of the 17th inst.

In reply thereto as well as to your former letter of the 16th inst., I am instructed to say that without either admitting or disputing the claim of the Anglo American Telegraph Company, to prevent him from carrying on the operations in which he has been engaged my client will, at least for the present, discontinue these operations and not resume them without first giving you or your clients sufficient notice to enable them to re-assert their claim.

With regard to your suggestion as to proposals with a view to the continuance of his work, my client is not on (sic) a position, until after conference with those who are associated with him in this enterprise, to make any proposal with a view to further operations in this country.

Yours truly,

(Sgd.) J. S. Winter.

Marconi continues in his report to his Directors by saying:

"His Excellency, the Governor Sir Cavendish Boyle, Sir Robert
Bond the Premier, and all the members of the Government
have done their best to help me in every way within their power,
and have intimated to me that, should I ask them officially, they
would consider the proposal for them to undertake all liability
towards the 'Anglo' in the event of our erecting a permanent
station here. I enclose press cuttings of the speeches delivered
at a luncheon which the Governor gave in my honour."

"On the 20th inst., I received from Ottawa, the following
telegram from Mr W. S. Fielding, Minister of Finance, Canadian
Government. 'Much pleased to learn that you contemplated
coming to Nova Scotia to continue your experiments in
wireless telegraphy. I assure you of a cordial welcome, and
the co-operation of any of the Government officials whose
knowledge would be useful to you. There are no difficulties
whatever in the way of your carrying out your operations there.'"

Marconi then gave full information to the Press. In Newfoundland
the arbitrary action of the Cable Company aroused much criticism
and produced a strong reaction favourable to him. The St John's
Municipal Council at its meeting on the 20th recorded in its Minutes
an account of Marconi's success in bridging the Atlantic by wireless on
the 12th December, and deplored the action of the Anglo-American
Telegraph Company.

The news was given wide publicity in New York and in the Canadian
papers, and among many messages of support Marconi received was one
from Alexander Graham Bell offering him the use of land at Cape Breton,
Nova Scotia. This he decided to inspect.

In fact the reaction of the Canadian and American Governments
and of the general public in these countries to the threat of the

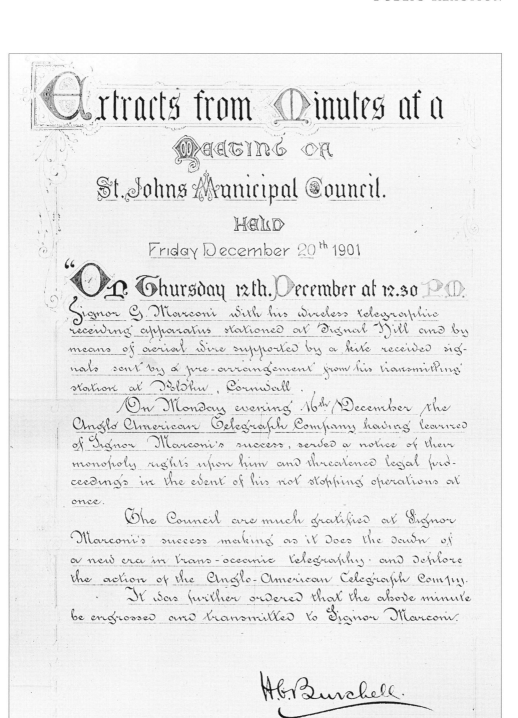

Extracts from the Minutes of St. John's Municipal Council,
20th December 1901

Cable from Alexander Graham Bell to Marconi, 18th December 1901,
and his reply

THE GREAT NORTH WESTERN TELEGRAPH COMPANY OF CANADA.

ALL MESSAGES TAKEN BY THIS COMPANY ARE SUBJECT TO THE FOLLOWING TERMS:

It is agreed between the sender of the following message and this Company that said Company shall not be liable for damages arising from failure to transmit or deliver, or for any error in the transmission or delivery of an unrepeated telegram, whether happening from negligence of its servants or otherwise, or for delays from interruptions in the working of its lines, for errors in cypher or obscure messages, or for errors from illegible writing beyond the amount received for sending the same.

To guard against errors, the Company will repeat back any telegram for an extra payment of one-half the regular rate, and in that case it shall not be liable for damages beyond fifty times the amount received for sending and repeating.

Correctness in the transmission of messages can be insured by contract in writing, stating agreed amount of risk, and payment of premium thereon at the following rates, in addition to the usual charges for repeated messages, viz.: one per cent. for any distance not exceeding 1000 miles, and two per cent. for any greater distance.

This Company shall not be liable for the act or omission of any other Company, but will endeavor to forward the telegram by any other Telegraph Company necessary to reaching its destination, but only as the agent of the sender and without liability therefor. The Company shall not be responsible for messages until the same are presented and accepted at one of its transmitting offices ; if a message is sent to such office by one of the Company's messengers, he acts for that purpose as the sender's agent ; if by telephone, the person receiving the message acts therein as agent of the sender, being authorised to assent to these conditions for the sender. This Company shall not be liable in any case for damages, unless the same be claimed, in writing, within sixty days after receipt of the telegram for transmission.

No employee of the Company shall vary the foregoing.

H. P. DWIGHT, President and General Manager.

Direct Connection with **ATLANTIC CABLES** and **WESTERN UNION** Telegraph Co.

SENT NO.	SENT BY.	REC'D BY.	TIME SENT.	TIME FILED.	CHECK.
☞ PLEASE WRITE PLAINLY.					

Send the following Message, subject to the above Terms, which are hereby agreed to.

190

To D'Alexander Graham Bell.
GIVE SUFFICIENT ADDRESS
Washington

Thanks for very kind telegram and invitation. Am not yet quite certain as to whether I shall be able to come to Washington. Shall telegraph again.

G. Marconi

261

☞ READ THE NOTICE AND AGREEMENT AT THE TOP.

61

Cable from Major Flood-Page to Marconi, 17th December 1901

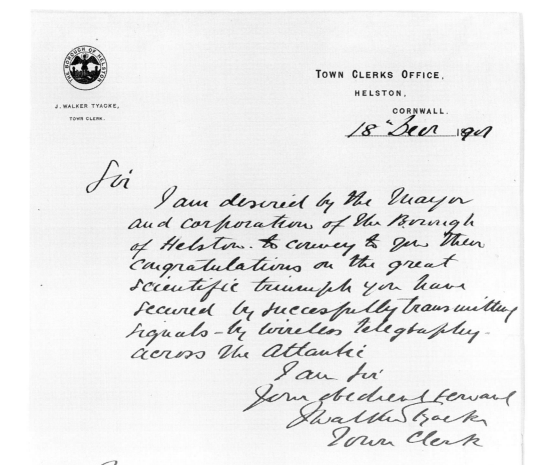

Letter to Marconi from J Walker Tyacke, the Town Clerk of Helston, 18th December 1901

Anglo-American Telegraph Company, furthered his future plans far more than would have resulted from demonstrating to interested parties the continuation of the occasional weak signals, almost masked by the crackle of atmospherics, that the apparatus then available allowed him to show.

The SS *Sardinian* docked at St John's on the 23rd December, and Paget with all his equipment and apparatus embarked the following day for England, having a research programme for a new receiver specified by Marconi to work out before the latter's return. Leaving St John's on the 24th, Marconi and Kemp spent Christmas Day travelling in a private railway coach, which belonged to Mr Reid the owner of the Newfoundland railway. That night they spent crossing to North Sydney, Nova Scotia, in a blizzard on board the SS *Bruce* belonging to R & G Reid of Montreal. Here they were welcomed by the Premier of Nova Scotia and members of the Government. They travelled later to Sydney and Glace Bay and on the 27th, despite the foul weather, they examined the neighbouring headlands by land and sea for possible sites.

They arrived in Ottawa on the 30th December and there followed a series of interviews, official receptions and entertainment. This demonstrated in no uncertain manner that Lord Minto, the Governor-General of Canada, Sir Wilfred Laurier, the Premier, the Minister of Marine, the Minister of Finance, and without exception every member of the Government they contacted, were at one in regarding Marconi as a benefactor of the seafaring community. They also agreed that his courageous enterprise in attempting to communicate across the Atlantic should be encouraged and given the utmost support possible.

The Government of Canada not only at once offered a free site, but also undertook to contribute $80,000 to the cost of building and equipping a wireless station at Cape Breton, provided the operating Company put a limit of 10 cents per word on commercial messages. This was fixed at 60% below the existing cablegram rate of 25 cents per word. Government messages and messages for the Press were to be transmitted at a ship-to-shore rate not exceeding 5 cents per word. The Government reserved the right to use the system for coastal signalling. The rate of exchange at the time was $4.80 to the £1 so the Canadian Government was contributing the equivalent of just over £16,000. When the final rate for messages was announced by the Company nearly a year later in sterling, private messages for England were accepted at 5 pence per word, which was 7½ pence less than the existing cable rates, in return for the Canadian Government subsidy. On the other hand Government and Press messages were to be charged at only 2½ pence per word.

A draft agreement was entered into between Marconi, acting for the Company, and the Canadian Government. Marconi left Ottawa on the 9th January 1902, for Montreal, the United States, and ultimately England. He was to return from there in February with the contract in its final form.

The day after Marconi and Kemp arrived at Montreal, on the 10th January, they had lunch at the St James Club with the Executive Staff of the Association of Electrical Engineers of Canada. They left Montreal for New York on the following evening by the New York Central & Hudson River Railway in a private stateroom at the end of a sleeping car. Kemp in his diary describes the scene in the morning: "We drew near to the River at daybreak and arose from our beds to admire the scenery surrounding it – it was certainly a grand sight; the River was frozen over and, as there had been a fall of snow during the night, there was a covering on the ice."

They arrived in New York Central Station at 9.00pm on the 12th January. The next day a quite exceptional dinner was given in Marconi's honour by the American Institute of Electrical Engineers in the Astor Gallery at the Waldorf-Astoria in New York. The affair had had to be arranged at virtually forty-eight hours notice owing to the uncertainty of Marconi's departure from Canada and his intention of sailing almost immediately from New York.

The dinner was a brilliant success. It was attended by members of the Institute and their ladies, a total of over 300. These included many people that Marconi would be pleased to meet such as Professors Steinmetz, Elihu Thomson, Alexander Graham Bell, and also Dr M I Pupin, Frank J Sprague and others.

It had all been very imaginatively arranged and it is described in the *Electrical World and Engineer*. The menu, printed in Italian olive ink on sea foam green card, bore outside a sketch representing an ocean scene with the two coasts of the Atlantic and a Marconi staff on two tall lighthouses signalling 'S' in three dots all the way across. In the middle was a medallion with Marconi's portrait draped with the Italian flag, above it the Italian coat of arms and below it, as a pendant, the Institute badge. At one side on the left was the American flag and on the right the British. It is said that the whole thing was designed, sketched, engraved, printed, and delivered within thirty-six hours.

At the two ends of the room were large tablets, one reading 'Poldhu' in white lamps and the other 'St John's', in letters about a foot long. Immediately opposite the speaker's table was a similar tablet bearing the name of 'Marconi'. Between the three signs were strung, at frequent intervals, clusters of three lamps to represent the three dots, or 'S', sent across the Atlantic from the Cornish coast to Newfoundland. At fitting times these were flashed or allowed to stay illuminated.

It was all an amazing tribute to Marconi and a very quick recognition of the importance of his achievement.

Looking back from the standpoint of present knowledge on the conditions prevailing during those momentous tests, we have a measure of the risk of failure that faced Marconi after his patient long-term preparations, and of the driving force that at this stage of his career impelled him bravely and successfully to grapple with so many technical obstacles of a magnitude that a professional physicist of that period would not have attempted to solve without several more years spent on preliminary study and experiment.

All this is illustrated by considering the stages in which Marconi's work

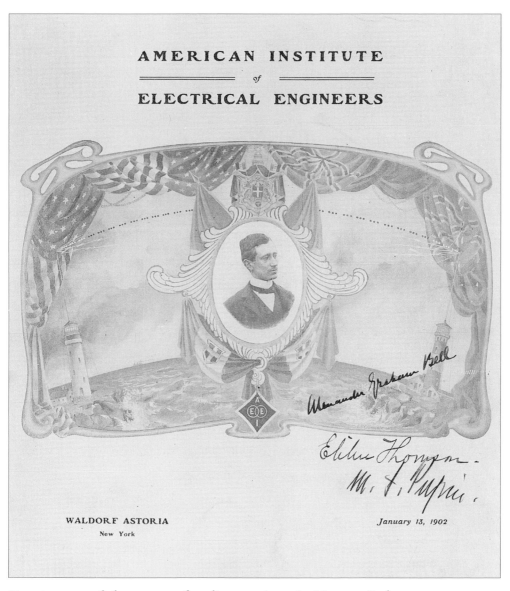

Front cover of the menu of a dinner given in Marconi's honour,
13th January 1902

developed. At Poldhu the input power was 15 kilowatts. No convenient method of measuring station wavelengths had been evolved at that time, but the circuit constants of the Poldhu transmitter indicated that, as stated later by Marconi in 1908, it was about 366 metres (this means a quarter-wavelength would be 300 feet). The syntonic tuning method of reception had been tested at Crookhaven, 225 miles from Poldhu, and had given excellent results. The height of the mast at Crookhaven was 150 feet and length of aerial wire possibly no more that 180 feet (at which it would have been a capacitive source).

At St John's a similar arrangement could have been obtained in fair weather by the use of a balloon or kite, but in the wind then prevailing, obtaining a similar height meant using an aerial appreciably longer than 180 feet. On the 12th December the first kite employed a pair of wires over 500 feet long, and the second kite a single wire of 500 feet in length. On each occasion the aerial could have been inductive and could have required a different tuning arrangement from that used at Crookhaven. Also the reactance of the aerial would have changed with the rise and fall of the kite. For both reasons the apparently untuned receiver with the self-restoring coherer used by Marconi gave better results.

As it happened, Marconi had departed from the usual method of connecting this mercury coherer direct in the aerial circuit. He mentions quite casually in one of his papers that he had placed it in the secondary circuit of an air-cored transformer designed to build up the potential across the coherer. It is possible that this transformer circuit may have contributed some measure of tuning effect which gave a more suitable adjustment for receiving, particularly on the long kite aerial, than did the syntonic tuned receiver adjusted to work on a shorter aerial with the normal coherer.

In the days after the dinner Marconi inspected the transatlantic station at South Wellfleet, Cape Cod, which by now was preparing for the construction of the four 210-feet wooden towers that were to be erected in place of the ring of 20 masts that had collapsed. The transmitting equipment similar to that at Poldhu had already been fully restored. He instructed Vyvyan to return with him to England for the purpose of assisting in the design, and later the construction, of the transatlantic station to be erected in Canada.

Preparations were being made during this time for Marconi's return to England. W W Bradfield, in association with Messrs Moeran and Bottomley, was managing the ship-fitting business in New York for the Marconi Company, and Kemp was provided with apparatus and given every assistance to equip the American Line SS *Philadelphia* with syntonic tuned wireless. On the 22nd January 1902, Marconi, Vyvyan and Kemp sailed on that ship from New York to Southampton.

During the voyage normal communication was established with the stations at Nantucket (USA), the Lizard, Niton and the Haven. On the 30th January Marconi disembarked at Southampton and was able to attend the Fifth Ordinary General Meeting of Marconi's Wireless Telegraph Company held on the 20th February. Marconi gave the Directors and shareholders a first-hand account of the Newfoundland tests and the reason for their abrupt termination. He obtained their whole-hearted endorsement of his actions. During this visit to England he had time to conduct further tuning experiments with signals between Poldhu, the Lizard, the Haven and Niton. The Marconi Company was also able to put the draft agreement with the Canadian Government into its final form.

SS Philadelphia, *February 1902*

Marconi's wireless cabin on board the SS Philadelphia, *February 1902*

Final Proof

The SS *Philadelphia* was due to leave Southampton again for New York on the 22nd February and Marconi decided to travel by this ship on his return journey, bringing with him the Agreement with Ottawa. At the same time, now that everyone knew what he planned to do in Canada, there was nothing to lose and possibly much to gain, if he openly carried out a range test over the lengthening Atlantic course between the SS *Philadelphia* steaming westward and Poldhu.

The constants of the oscillating circuit at Poldhu would be the same as for the Newfoundland tests, the wavelength radiated being as before, about 366 metres. However, the keying of the transmitter had been improved so that sequences of dashes as well as dots, and therefore a variety of messages, could be sent.

The conditions of reception also had the great advantage of a stable aerial support, provided by the ship's masts, and an excellent earth connection via the ship's hull. The successful issue of such tests was never for a moment in doubt as far as Marconi himself was concerned. However, apart from acquiring more useful experimental data, he felt that by taking the earliest opportunity to confirm by records, for all to see, that signals did somehow cross the width of the Atlantic, he could remove any lingering doubts his Canadian friends might have on the subject and greatly encourage their further co-operation. The American Line had agreed to extensions to the main and mizen masts. Here is Kemp's description: "I fitted a 4 part aerial on three bamboo crosses made with 14 ft. bamboos; one at the main mast head, the second at the mizen mast head and the third between the two nearest boat davits to the Instrument room." The Poldhu receiving equipment was installed in a special cabin, one of a suite of four cabins on the upper deck aft. So equipped, the SS *Philadelphia* sailed from Southampton on the 22nd February 1902, with Marconi and his party. This included H S Saunders, a Director of the Company, Vyvyan and J D Taylor, the engineers for the proposed Canadian station, while F S Stacey and C S Franklin operated the ship-to-shore apparatus.

Communication with Niton was maintained up to the usual 70 miles and Marconi then concentrated on the reception of signals from Poldhu. As attested by Captain Miles of the SS *Philadelphia*, readable messages were recorded during daylight up to 700 miles, and after darkness set in, up to 1551 miles. The last message received at this distance on the 25th February ticked out on the Morse inker tape: "All in order. Do you understand?" More importantly, the signal 'S' came through distinctly up to 2099 miles using a filings coherer. About half these ranges were obtained by employing the self-restoring mercury coherer.

However, when it is remembered that the aerial on this occasion had only 150 feet of mast height, whereas at Signal Hill the kite-supported aerial had a length of 500 feet and was nearly vertical, and that allowance must be made for a difference in tuning arrangements on the two occasions, these results are consistent with the reception at Signal Hill. The hostile critics of Marconi's statements about the Newfoundland test were silenced. He had won through on this fresh evidence. This was certainly the opinion of the general public, for in the Spring of 1902 the value of cable shares on the London Stock Exchange moved downwards.

Curiously enough, Lord Londonderry, Postmaster General, objected that these transatlantic experiments might interfere with the experiments of the Admiralty. And he did this at a time when the Royal Navy was using the

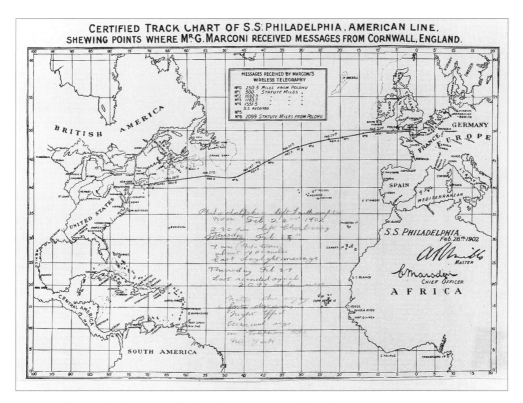

Track chart of the SS Philadelphia *certified by the Captain,*
28th February 1902

Company's apparatus and the Treasury was paying the Company £5000 per annum.

For centuries man had sailed the seas, but at no period in history, whatever the speed and luxury of sea travel, had it ever been possible for a ship over the horizon to speak with the outside world. Every ship that sailed became an isolated world completely cut off from all contacts beyond the community of the ship. How magical, then, must it have seemed to the passengers aboard the SS *Philadelphia* to find at their disposal a means of sending and receiving messages from friends 1500 miles distant.

On the purely scientific plane this voyage on the SS *Philadelphia* was of great importance as it enabled Marconi to discover another fundamental characteristic of the signal-conveying medium, namely that messages travelled far further by night than by day. As regards the marine use of wireless the voyage had also shown that, with a station similar to Poldhu on both sides of the North Atlantic, it should be possible for a vessel to keep in touch with land the whole way across. To reporters in New York, Marconi prophesied that one day a wireless message would be sent right round the world, being received, after the calculated brief interval, at its place of origin. This was actually accomplished some 24 years later.

After landing in New York from the SS *Philadelphia*, Marconi proceeded to Ottawa with the contract in its final form. The Agreement was duly ratified by the Dominion House of Commons, when Lord Minto paid a fine tribute to Marconi. Then Marconi, accompanied by Vyvyan, went to Cape Breton Island and, after inspecting many alternative sites for the transatlantic station, finally chose one at Table Head, Glace Bay. Vyvyan was detailed to put in hand the plans for the station and was appointed Managing Engineer for the Marconi Company of Canada. Marconi returned to New York and while there the Marconi Wireless Telegraph Company of America, which had been registered on the 22nd November 1899, was incorporated on the 1st April 1902 as a public company. The American rights in Marconi inventions were transferred to the new organisation by the Parent Company for the sum of £50,000 with shares giving a controlling interest in its operation. Bradfield, whose service with the Parent Company dated back to 1897, became Chief Engineer of the American company.

Towards the end of 1902 Marconi had a permanent station on the east coast of Canada at Glace Bay, and tests were started with the stations at Poldhu and Cape Cod.

Details of the quantities of wood for the new wooden towers at Poldhu, January 1902 (mistakenly dated 1901)

Quantities for Poldue Tower.

Particulars	No	length	width	thick	Descrip tion
Main leg					
Sleepers at base	16	10'.0	12"	6"	PP
Vertical posts	128	31'.0	8"	8"	PP
do	24	31'.0	8"	8"	PP
Diagonal bracing to V posts	432	13'.0	7"	2½"	RD
Hoz do	80	6'.0	7"	2½"	RD
Horizontal bracing girders	48	41'.0	8"	8"	PP
Bracing to do	144	9'.0	7"	2½"	RD
Diagonal bracing ties	6000 ft run		7"	2½"	RD
in 12'.0 to 18'.0 lengths					
Diagonal bracing girders top	6	31'.0	8"	8"	PP
do	2	20'.0	8"	8"	PP
do	8	17.0	8"	8"	PP
do	4	26.0	8"	8"	PP
Bracing to do.	28	9'.0	7"	2½"	RD
Brackets at top	8	31'.0	8"	8"	PP
bracing to do	20	12'.0	9"	3"	RD
Beams at top	4	45.0	12"	6"	PP

P.P. - Pitch pine
RD - Red deals.

PHS Jan 30 1905.

New wooden towers at Poldhu under construction, Spring 1902

Above: *The completed towers at Poldhu, Autumn 1902*
Left: *A rare photograph taken from a balloon, of the Poldhu Hotel and one tower, Spring 1903*

Messages from important people soon began to flow before the end of the year, when good operating periods for transmissions prevailed. These were reported in the newspapers as follows.

> The Governor-General of Canada, Lord Minto, to the King
> of England:
> "To His Majesty the King, London. May I be permitted by means of this wireless message to congratulate your Majesty on success of Marconi's great invention connecting England and Canada. Minto."

> Marconi to Lord Knollys for the King:
> "To Lord Knollys, Buckingham Palace, London. On occasion of first wireless telegraphic communication across Atlantic Ocean may I be permitted to present by means of this wireless message transmitted from Canada to England my respectful homage to His Majesty the King. Marconi."

> Marconi in Italian to the King of Italy via General Brusati, his Aide-de-Camp:
> "Occasione prima transmissione radiotelegrafica Transatlantica invio con questo telegramma trasmesso attraverso lo Spazio dal Nuovo al Vecchio Mondo devoti ommagge a sua maesta il Re. Guglielmo Marconi."
> *Translation:* "(On the) occasion (of the) first transmission (by means of) Transatlantic radiotelegraphy, I send with this telegram, transmitted across the space from the New to the Old World, devoted homage to His Majesty the King. Guglielmo Marconi."

All three messages were released from Poldhu and forwarded by land-line on the 21st December and replies were soon to follow.

Marconi's wireless station at Glace Bay, showing the wooden towers during construction, May 1902

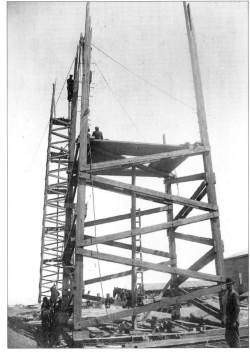

Canada to England. Dec: 5th 1902. First tape signals received.

Tape showing message in Morse code, 5th December 1902, and the coherer receiver (right in picture above) on which it was received at Poldhu. To the left in the picture is to be seen an early magnetic detector

Opposite above: *The completed wireless station at Glace Bay, December 1902*

Opposite below: *Marconi and staff at Glace Bay, December 1902*

JACK THE GIANT KILLER.

Cartoon, Western Mail , December 1902

His Majesty King Edward to Lord Minto:

"I am much interested by the wireless message which you have sent me, and I am delighted at the success of Signor Marconi's great invention, which brings Great Britain and Canada into still closer communication. Edward."

Lord Knollys to Marconi, Canada:

"I have had the honour of submitting your telegram to the King, and I am commanded to congratulate you sincerely from His Majesty on the successful issue of your endeavours to develop your most important invention. The King has been much interested by your experiments, as he remembers that the initial ones were commenced by you from the Royal yacht *Osborne* in 1898. Knollys."

King Victor Emmanuel to Signor Marconi:

"The King learns with lively satisfaction of the splendid result, constituting a new and glorious triumph to Italian science. Victor Emmanuel."

Late on the 18th January 1903, Cape Cod transmitted the first wireless message to be received in England direct from the United States.

President of the United States to the King of England:

"To His Majesty King Edward the Seventh. In taking advantage of the wonderful triumph of scientific research and ingenuity which has been achieved in perfecting the system of wireless telegraphy, I extend on behalf of the American people my most cordial greetings and good wishes to you and the people of the British Empire. Theodore Roosevelt."

King of England to the President of the United States (19th):
"To the President, White House, Washington. I thank you most
sincerely for the kind message which I have just received from
you through Signor Marconi's transatlantic wireless telegraphy.
I sincerely reciprocate, in the name of the people of the British
Empire, the cordial greetings and friendly sentiment expressed by
you on behalf of the American nation, and I heartily wish you and
your country every possible prosperity. Edward R."

This final reply was sent, apparently, on the evening of the 19th by
cable, as the telegraph office at Mullion, Cornwall (to which all messages
intended for the Company's long-distance station had to be transmitted)
was closed for the night.

A trickle of messages continued to pass from station to station, but it was
clear by this time that further development work was needed. It was to be
another five years before a full commercial transatlantic wireless service
became established.

In July 1903, Marconi received the Prince and Princess of Wales at
Poldhu where they paid a very happy visit. After the party of 15 had lunched
at the Poldhu Hotel, they visited the wireless station and displayed the
greatest interest in everything that was shown to them. The Prince, who
knew the Morse code, witnessed the reception of messages from the
wireless station at the Lizard. He, a Maid-in-Waiting, and a few others
actually climbed to the top of one of the towers where there was an
awning and a telescope that enabled the party to secure extensive views.
However, the Princess gave up the attempt at the second stage and
returned to terra firma. Later in the afternoon the Royal party motored
to the Lizard and walked along the cliffs before taking tea with Marconi in
the Housel Bay Hotel.

Poldhu Hotel and wireless station towers decorated for the Royal visit

The Royal party pictured arriving at Poldhu, 18th July 1903

Marconi on the steps of one of the towers at the time of the Royal visit to Poldhu

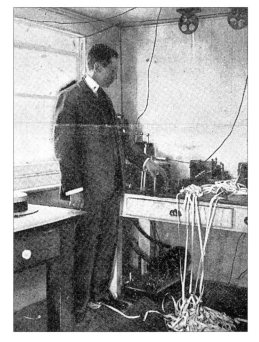

The Morse code tapes with messages from the Lizard wireless station emerging during the Royal visit from a Morse inker at Poldhu

Opposite: *The Prince and Princess of Wales, and the Royal party, on their tour of the wireless station at Poldhu with Marconi as their guide*

Marconi's Success

This has been the story of how Marconi bridged the Atlantic by wireless. It had seemed to many people an impossible feat. Marconi's achievement had been tremendous.

Lord Rayleigh, who was originally sceptical, was one of a distinguished company of scientific doubters who were later as generous in their tributes as they had been forthright in earlier criticism. In a lecture two years later, when the transmission of transatlantic wireless signals was an established fact, Lord Rayleigh is reported to have said:

> "I have known Mr Marconi for some time and have never known him to say he could produce results unless he was sure he was able to do so. I would have thought the curvature of the earth would have been a drawback in signalling from England to America, for there is a mountain of water one hundred miles high to be got over, and electric waves tend to go straight. The fact that the waves are bent round this mountain is an illustration of the phenomenon of diffraction."

The reason why the wireless waves were able to overcome the obstacle of the earth's curvature was not discovered until later.

The history of science has many great moments when success has come as reward, sometimes for an illuminating flash of inspiration, but more often for faith and work. It was no easy conquest, that of the Atlantic. It did not merely involve the solution of laboratory problems. It imposed physical hardships and hazards, disappointments and setbacks, and these required the moral qualities of fortitude, courage and steadfastness in the face of apparently overwhelming odds. Marconi achieved his success in less than two years, in a world that had cable telegraphy but no air travel, few cars, and limited telephone communications. He combined his scientific knowledge with his exceptional drive and business acumen, and what he accomplished in those first two years of the 20th Century would affect mankind for the rest of time.

Sources

At Marconi plc Archives

Marconi, G: letters, reports and lecture scripts (1900-1903)

Fleming, A: letters, reports and lecture scripts (1900-1903)

Kemp, G: letters and diary entries (1900-1903)

Flood-Page, S: letters (1900-1902)

Miscellaneous: company documents and legal agreements, unpublished personal records

At BBC Sound Archives

Marconi and Paget (1935) recalling transatlantic wireless in 1901

At The Public Record Office

Ships' Passenger Embarkation Lists (1900-1902)

Newspapers and Periodicals

' *The Daily Express* (1902)

The Financial Times (1902)

The Daily Telegraph (1901-1902)

The Times (1901-1903)

The Engineer (1895)

Electrical World and Engineer (1902)

The Electrician (1902)

The Weekly Marconigram (1903)

Books

Clerk, D	*The Gas and Oil Engine* (Longmans, Green, and Co. 6th edition revised and enlarged 1896)
Fleming, A	*The Principles of Electric Wave Telegraphy* (Longmans, Green, and Co. 1906)
Beaumont, W W	*Motor vehicles and motors: their design construction and working by steam, oil and electricity* (Constable and Co. 2nd edition revised 1902)

Other works by the author

Gordon Bussey   *Vintage Crystal Sets 1922 -1927*
IPC Electronic-Electrical Press (1976)

Gordon Bussey   *The Story of Pye Wireless*
Pye Limited (1976)
2nd edition (1983)
3rd edition (1986)

Gordon Bussey   see entries in *Dictionary of Business Biography*
Butterworths 5 vols.
Vol. I E K Cole (1985)
Vol. IV H J Pye (1985)
Vol. IV W G Pye (1985)

Gordon Bussey   *Wireless: the crucial decade 1924 - 34*
Peter Peregrinus Ltd (1990)

Gordon Bussey   *The first use of wireless by the Royal family 1898*
Issued for a commemorative re-enactment, 1998
The General Electric Company (1998)

Gordon Bussey   *Marconi Centenaries in 1999*
Marconi Communications (1999)
2nd edition (1999)

Keith Geddes   *Television: the first fifty years*
& Gordon Bussey   National Museum of Photography, Film &
Television (1986)

Keith Geddes &   *The History of Roberts Radio*
Gordon Bussey   Roberts Radio Co. Ltd (1988)

Keith Geddes in collaboration with Gordon Bussey	*The Setmakers* The British Radio & Electronic Manufacturers Association (1991)
Geoffrey Dixon-Nuttall & Gordon Bussey	*Hacker Radio* GDN Publications (1993)
Gordon Bussey (Ed.)	Supplement to V & A Exhibition, *The Wireless Show!* Philips Industries (1977)
Gordon Bussey (Ed.)	Supplement to Science Museum Exhibition, *The Great Optical Illusion* Philips Industries (1980)

Index

Note: abbreviation 'M' means (in this index) 'Marconi'